Energy 107

利用太阳能的蓝龙

The Solar Blue Dragon

Gunter Pauli

[比] 冈特·鲍利 著

[哥伦] 凯瑟琳娜·巴赫 绘

高 芳 李原原 译

上海远东出版社

丛书编委会

主　任：田成川

副主任：何家振　闫世东　林　玉

委　员：李原原　翟致信　靳增江　史国鹏　梁雅丽
　　　　任泽林　陈　卫　薛　梅　王　岢　郑循如
　　　　彭　勇　王梦雨

特别感谢以下热心人士对童书工作的支持：

匡志强　宋小华　解　东　厉　云　李　婧　庞英元
李　阳　刘　丹　冯家宝　熊彩虹　罗淑怡　旷　婉
杨　荣　刘学振　何圣霖　廖清州　谭燕宁　王　征
李　杰　韦小宏　欧　亮　陈强林　陈　果　寿颖慧
罗　佳　傅　俊　白永喆　戴　虹

目录

Contents

一群蓝瓶僧帽水母沿着海岸游动时，遇到了一条孤独的沙丁鱼。

A school of bluebottles is floating along the coast when they meet a lonely sardine.

蓝瓶僧帽水母遇到了一条孤独的沙丁鱼

Bluebottles meet a lonely sardine

过度捕捞导致我们快要灭绝

Overfishing has really wiped us out

"哎呀，你是这儿为数不多的沙丁鱼之一了。"僧帽水母说。

"是的，过度捕捞导致我们快要灭绝了。你一定很高兴渔民们帮你摆脱了最大的食物竞争对手。"沙丁鱼回答。

"Oh, you are one of the few still around," says the bluebottle.

"Yes, overfishing has really wiped us out. You must be happy that the fishermen got rid of your biggest competitor for food," replies the sardine.

"我知道。但还有可怕的带须虾虎鱼，他真的很喜欢吃我们！"

"你应该也害怕蓝龙吧！"

"I know. But there is still that dreaded bearded goby fish and he really likes to eat us!"

"You should also be scared of the blue dragon!"

可怕的带须虾虎鱼

Dreaded bearded goby fish

这个蓝龙会喷火焰

This blue dragon spews fire

"又是龙！"僧帽水母喊道。"这些天，人人都在谈论龙。"

"这个蓝龙会喷火焰。"

"哦，是的。我太了解这一切了！他利用我们制造他的火！"

"Not dragons again!" shouts the bluebottle. "These days, everyone talks about dragons."

"This blue dragon spews fire."

"Oh yes, I know that all too well! He makes his fire from us!"

"你一定是
在开玩笑！蓝龙怎
么会用蓝瓶僧帽水母来制造
火呢？"

"好吧，你知道我们的毒素能让人感
到火辣辣的痛。"

"我当然知道，那就是被你刺
到会很痛的原因。"

"You must be joking!
How does the blue dragon make
fire from a bluebottle?"

"Well, you know that we have this
toxin that burns."

"Of course I do; that's what
makes your sting so painful."

那就是被你刺到会很痛的原因

That's what makes your sting so painful

像火烧一样疼

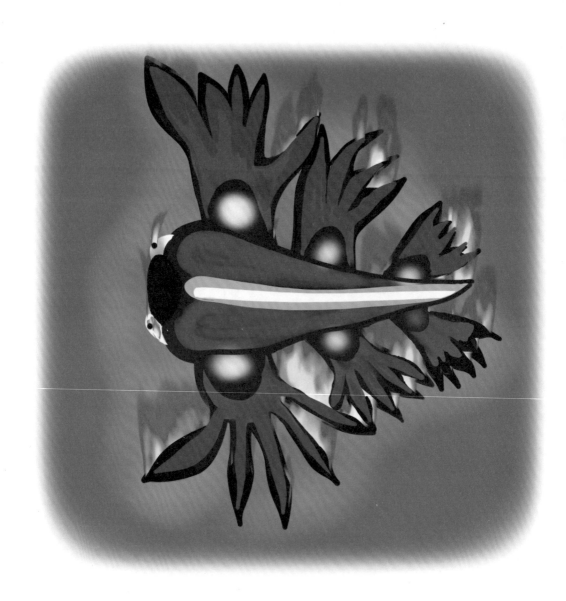

It really burns like fire

"没错。所以，蓝龙吃掉我们，把我们所有的毒液放进一个特殊的小口袋，当他受到威胁时再吐出来。这些毒液让人感到像火烧一样疼。"

"那蓝龙是如何摆脱捕食者的呢？"

"Exactly. So, the blue dragon eats us, puts all our venom into a special little pocket, and spits it out when he is threatened. And it really burns like fire."

"How does this blue dragon get away with it?"

"正如他的名字所示——他是蓝色的。他漂浮在水面上，蓝色的肚子朝上，这使鸟类在俯瞰蓝色的海洋时，很难看到他。而且他有闪亮的灰色后背，下面的鱼从底部向上看时，也看不到他。"

"但你知道吗？他其实是一个软体动物，只是在行为上更像一条龙。"

"As the name says – he is blue. He floats on the water with his blue tummy upwards, which makes him hard for birds to see when they look down on the blue ocean. And his shiny grey back that is seen from the bottom makes him invisible to fish below."

"But did you know that he is in fact a mollusc and only behaves like a dragon?"

正如他的名字所示——他是蓝色的

As the name says - he is blue

他确实是一个没有壳的贝类

He is a shellfish without a shell

"软体动物，也就是说，他是个贝类动物？可是他都没有壳。"

"我知道这很让人费解，但他确实是一个没有壳的贝类。"

"这还不算什么，有的蓝龙是靠太阳能生存的。"

"不是吧，你一定是在开玩笑！一条利用太阳能的龙？"

"A mollusc, in other words a shellfish? But he doesn't even have a shell."

"I know it's confusing. He's a shellfish without a shell."

"And that is only the beginning. There are blue dragons that live on solar energy."

"No, you must be kidding me! A solar dragon?"

"虽然不可思议，但那是事实。他让微小的藻类生长在他的身体里，供给他所需的糖，这样他可以靠阳光生存，而不用吃东西。我们僧帽水母，只不过是他的甜点。"

"对我来说，他现在听起来才像一条真正的龙！"

……这仅仅是开始！……

"Incredible but true. He has these tiny algae growing in his body that give him sugars so that he can live on sunlight instead of real food. We, the bluebottles, are only his dessert."

"Now that sounds like a real dragon to me!"

... AND IT HAS ONLY JUST BEGUN!...

······这仅仅是开始！······

... AND IT HAS ONLY JUST BEGUN! ...

蓝龙有很多名字：海燕子、蓝色天使以及蓝色海蛞蝓。

The blue dragon has many names: sea swallow, blue angel, and blue sea slug.

蓝龙靠吞咽空气并将其保存在胃旁边的一个袋内来漂浮，这个空气口袋让蓝龙能够肚皮朝上漂浮。

The blue dragon floats by swallowing air and keeping it in a pouch that is next to its stomach. The pocket of air enables the dragon to float upside down.

小小的蓝龙不产生任何毒液，但它从猎物那里获取毒液，集中储存在触须上，这样更致命。

The tiny blue dragon does not make any venom, but it takes venom from its prey, concentrates it, and stores it in its tentacles, making it more lethal.

蓝龙是一个有攻击性的捕食者，以比自己大得多的生物（包括僧帽水母）为食。蓝龙会吃掉僧帽水母那50米长的有毒触须。

The blue dragon is an aggressive predator that feeds on organisms that are much larger than itself, including the Portuguese man of war, devouring its toxin laced 50 m long tentacles.

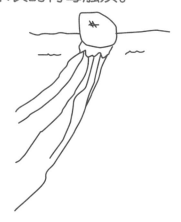

蓝龙舌头的形状就像锯子，这就是它可以迅速从猎物身上切下一块肉的原因。

The blue dragon has a tongue that is shaped like a saw. That is why it can cut off a piece of its prey very quickly.

蓝龙可以将阳光转化为糖。它们比人类更早利用太阳能进行光合作用。它们体内生长着微小的藻类，这些藻类在受保护的环境中繁殖旺盛。

Blue dragons can convert sunlight into sugars. They are ahead of humans in harnessing solar energy through photosynthesis. Some of the species capture and farm microscopic algae inside their bodies, where they flourish in a protected environment.

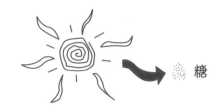

糖

沙丁鱼只吃浮游生物，这使它们富含营养且安全。

Sardines eat only plankton. That makes them rich in nutrients and safe to eat.

沙丁鱼没有武器来保护自己免受捕食。它们的防御机制是1 000万条以上的鱼一起成群游动。这会迷惑捕食者，让它们误以为自己看到的生物有5千米长、1千米宽。

Sardines have no weapons to protect themselves against predators. Their defence mechanism is to swim together in schools of up to 10 million fish. This confuses predators who is fooled into thinking they are seeing a creature that is 5 km long and 1 km wide.

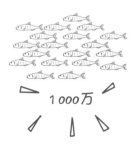

1 000万

Think about It

想一想

Can you imagine that a small creature of only few centimetres long could take on a large one with 50 m long tentacles?

你能想象，一个只有几厘米长的小生物可以挑战长着50米触须的对手吗？

如果你肚子朝上漂浮，你会怎样保护自己的背部？

If you were floating belly up, how would you protect your back?

Do you believe that firespewing dragons really exist in the water?

你相信水里真的存在喷火的龙吗？

对捕捞渔业的思考：我们应该留一些鱼在水中给海狮和鹈鹕吃吗？

Think about fishing: Should we leave some fish in the water for the sea lions and the pelicans to eat?

Explore history and look for examples where someone small took on someone big, and won. Take note of how they achieved the impossible to win the battle. Now look at Nature and find tiny predators that take on big prey, and succeed. Again, discover ways in which it is possible for a tiny predator to beat the odds and win. When you compare the examples from history and nature, draw some conclusions on how we can achieve the seemingly impossible.

　　探索历史，寻找以弱胜强的例子，注意他们如何在这场看似不可能赢的战斗中取胜。现在看看大自然，寻找小的捕食者战胜大的猎物的例子。然后，看看小的捕食者是如何获取出人意料的胜利的。对比历史和自然的例子后，得出关于如何实现看似不可能的目标的一些结论。

学科知识

Academic Knowledge

生物学	死的蓝龙和僧帽水母仍然可以投刺；像蓝龙一样，海龙偶尔也会同类相食；沙丁鱼在食物链中级别很低，因此在有汞和多氯联苯污染物的环境下非常少见；水母能够麻醉食肉动物的中枢神经系统和大脑；运动神经末梢调节鱼群的运动；海豚与沙丁鱼一起行动，把它们当成"诱饵球"。
化 学	蓝龙收集僧帽水母的刺丝囊及其毒素；沙丁鱼富含维生素B$_{12}$（支持神经系统），还有磷、钙、钾、硒、ω-3脂肪酸和维生素D，能增强对钙的吸收，对骨骼健康至关重要。
物 理	水母拥有一个复杂的视觉系统，为其在沼泽里导航；水母的外层感光器官帮助它区分明暗。
工程学	如何使物体在水中隐形；沙丁鱼油用于制造油漆和清漆；通过研究鱼群和鸟群的运动，交通管理员应该掌握汽车在繁忙的高速公路上如何移动会增加交通事故风险。
经济学	生产1千克三文鱼肉需要3千克饲料鱼，如果人们更愿意吃三文鱼，我们的需求就会增加为原先的3倍。
伦理学	野生捕捞的鱼被用作饲料后仅仅提供三分之一的蛋白质，我们如何证明用野生鱼类作为养殖鱼的饲料是合理的呢？
历 史	1777年在库克船长驾驶"决心号"进行的第二次太平洋航行里，蓝龙被科学家首次记录；拿破仑·波拿巴推广沙丁鱼是为了生产沙丁鱼罐头，史上第一个鱼罐头就是为了满足他所管理的市民和士兵的需求。
地 理	5月至7月之间，沙丁鱼在非洲南部海岸很常见，引起了大型海洋食肉动物（如海豚、大白鲨、铜鲨鱼、角塘鹅）和许多海鸟的聚集；蓝龙出没在南非海岸、莫桑比克、欧洲南部和澳大利亚；水母和僧帽水母用阳光作指南针为其导航。
数 学	大数定律是指在随机试验中,每次出现的结果不同,但是大量重复试验出现的结果的平均值却非常接近某个确定的值；"数字安全"理论就是每个单独的鱼在鱼群里被抓住的可能性变得相对较低。
生活方式	可持续的食物选择：植物性食物需要更少的能源和生产土地空间，产生较低的碳排放，只需本地养殖的植物即可。
社会学	用"挤得像沙丁鱼"来形容非常满的公共汽车或火车上的人；沙丁鱼成群行动，这是一种防御机制。
心理学	要有信心战胜远远大于自己的猎物，小的可以战胜大的。
系统论	当沙丁鱼被过度捕捞，哺乳的雌性海狮只能吃营养很少的食物，因此只能提供给幼崽更少的奶；沙丁鱼最近的数量增加了一倍，因为食肉动物物种，包括鲨鱼、金枪鱼、鳕鱼和海豚也被过度捕捞。

情感智慧
Emotional Intelligence

蓝瓶僧帽水母

蓝瓶僧帽水母很自信，但对孤独的沙丁鱼没有同情心。知道沙丁鱼受到过度捕捞的影响后，僧帽水母感到不安，听到"龙"这个词时，他感到心慌。僧帽水母厌倦了一遍又一遍地讨论相同的问题。不过，他还是告诉沙丁鱼，蓝龙的毒液来自猎物的毒液。僧帽水母难以说服沙丁鱼，这些微小的动物可以猎食比他们大的动物。僧帽水母花时间来解释蓝龙如何成功地在水中移动而不被注意。他对沙丁鱼的解释很感兴趣。

沙丁鱼

沙丁鱼艰难地忍受孤独，这是过度捕捞的结果，他已接受命运。沙丁鱼想要警告僧帽水母防备蓝龙，但得到了一个令他惊讶的答案。虽然在改变自己的生活条件上什么都没做，但他惊讶地了解到很多关于蓝龙的新信息。沙丁鱼折服于僧帽水母解释一切的能力，也折服于蓝龙独特的不引起注意的移动能力。当沙丁鱼得知蓝龙甚至可以利用太阳能时，他表示，如果这种生物真的存在，那就是他想象中的龙应有的样子。

艺术
The Arts

　　找找裸鳃类动物的照片,指出那些你觉得最不可思议的形状和颜色。看起来真的很惊人!把它们与著名的漫画和卡通人物的图片作对比，有什么相似的地方吗？哪些作家和艺术家的灵感来自蓝龙？

思维拓展
Systems: Making the Connections

　　沙丁鱼和其他鱼类的过度捕捞不仅导致那些几十年来依赖一种鱼类的持续供应的企业出现财务崩溃，而且也剥夺了其他海洋生物生存所需的营养。我们必须意识到，大量的野生沙丁鱼被用作养殖鱼类（如挪威三文鱼）的饲料，它们已被严重商业化。今天的消费者更喜欢三文鱼，并且在市场销售的三文鱼几乎都是养殖的。连那些在智利和挪威海岸捕捞到的三文鱼，也是从养鱼场逃出来的。

　　我们必须发展可持续的渔业和农业。捕一种鱼来养活其他鱼，这种方式很难满足世界上不断增长的人口及其需求。我们不能期望陆地和海洋生产更多东西，而是应该更有效地利用现有的东西。这意味着，我们不要吃运往世界各地的高成本的三文鱼，而应该吃更多的像沙丁鱼一样在本地就能捕获到的鱼。沙丁鱼被认为是世界上最有营养的动物产品，为什么我们要用密集养殖的三文鱼取代这些富含营养的深海鱼呢？据发现，供应只有中高层收入者能买得起的昂贵鱼类（如三文鱼和鳟鱼），企业更有利可图。然而，40%的人口买不起昂贵的鱼，只能找更便宜的替代品。

动手能力
Capacity to Implement

　　你和家人吃三文鱼吗？算算每千克三文鱼的成本，与沙丁鱼比较一下。和预期一样，三文鱼更加昂贵，因为生产1千克三文鱼肉需要消耗3千克的沙丁鱼。想办法告诉人们，用吃三文鱼的同样价格，可以吃2倍的沙丁鱼或鲱鱼，还能为海狮和海鸟在海洋中留下足够的食物。为什么海狮和海鸟的茁壮成长对你很重要？同你的朋友和家人分享所有你获得的信息，提出令人信服的观点。他们可能会觉得数据很无聊，告诉他们令人难以置信的蓝龙来激发他们的兴趣吧！

故事灵感来自
This Fable Is Inspired by

伊丽莎白·曼–博尔杰塞
Elisabeth Mann-Borgese

伊丽莎白·曼–博尔杰塞生于慕尼黑，是德国作家托马斯·曼最小的女儿。她搬到美国后，以编辑和作家为职业，最终晋升为《大英百科全书》执行秘书。事业中途，她将兴趣转移到海洋，于1970年组织了第一次海洋法会议，主题是"海洋和平"。她促使各国政府订立了《联合国海洋法公约》。在59岁，她成为达尔豪斯大学教授，在那里教学直到81岁。她的书《海洋的戏剧》是第一部关于过度捕捞和公海污染危机的作品。伊利莎白是罗马俱乐部成员，并教她的德国牧羊犬弹钢琴。

图书在版编目(CIP)数据

冈特生态童书.第三辑修订版:全36册:汉英对照 /
(比)冈特·鲍利著;(哥伦)凯瑟琳娜·巴赫绘;
何家振等译.—上海:上海远东出版社,2022
书名原文:Gunter's Fables
ISBN 978-7-5476-1850-9

Ⅰ.①冈… Ⅱ.①冈… ②凯… ③何… Ⅲ.①生态环
境-环境保护-儿童读物—汉、英 Ⅳ.①X171.1-49

中国版本图书馆CIP数据核字(2022)第163904号
著作权合同登记号图字09-2022-0637号

策　　划 张　蓉
责任编辑 祁东城
封面设计 魏　来 李　廉

冈特生态童书

利用太阳能的蓝龙

[比]冈特·鲍利　著
[哥伦]凯瑟琳娜·巴赫　绘
高　芳　李原原　译

记得要和身边的小朋友分享环保知识哦!
八喜冰淇淋祝你成为环保小使者!